Tropical Forests: Botanical Dynamics, Speciation, and Diversity.

Abstracts from the AAU 25th Anniversary Symposium.

Edited by
Flemming Skov and Anders Barfod

AAU REPORTS 18

Botanical Institute Aarhus University 1988

The symposium is sponsored by The Danish Natural
Science Research Council, The Research Foundation
of the University of Aarhus, and Aarhus Olie.

Introduction

This booklet contains abstracts of lectures and posters presented at the AAU 25th Anniversary Symposium: "Tropical forests: Botanical dynamics, speciation and diversity".
A wide range of subjects from taxonomy to computer science and geological history are covered. The taxonomic treatments include examples from tropical Asia, Africa, and South America and discuss the phylogeny and evolution of these groups.
Floras and forest inventories span from Andean montane vegetation and rain forest in the Amazon to the flora of Sri Lanka and south India. The dynamics of tropical forests is treated in the conceptual framework of eco-units and gap-phase dynamics. Breeding biology, pollination and tree architecture describe the organization and function of tropical plants.
Geological history, including the study of fluvial disturbance, is a new approach which contributes to our knowledge about vegetation dynamics. Examples are taken from the Amazon basin in South America.
Economic botany and ethnobotany are treated with examples from agroforestry and the study of new potential crops.
The history of botany, the use of computers in plant sciences, the art of botanical illustrations and many more topics are mentioned.

The abstracts are listed alphabetically by first author. A list of all symposium participants and their addresses may be found in the back of this book.

Flemming Skov

and

Anders Barfod

Acknowledgements

Botany in Thailand, Ecuador, and Denmark
25 years of collaboration - 1963-1988

The staff and students of the Botanical Institute, University of Aarhus (AAU), gratefully acknowledge the fruitful and cordial collaboration with institutions in Thailand and Ecuador during the last 25 years.

Royal Forest Department, Forest Herbarium (BKF), Bangkok
The main counterpart in the Flora of Thailand Project, The Forest Herbarium, has contributed considerably to the development of tropical botany. Joint expeditions have been carried out in Thailand and many Thai botanists have visited Aarhus.

Department of Agriculture, Technical Division, Botanical Section (BK), Bangkok
The BK-herbarium with its classic collections of Kerr, Garrett, and 20th century Thai collectors has been invaluable for the revisions for Flora of Thailand.

Chulalongkorn University, Department of Botany, Bangkok
Thailand's oldest university has throughout the years carried out taxonomic work relating to the Flora of Thailand and many students from Chulalongkorn University have visited Aarhus for training.

Prince of Songkla University (PSU), Department of Biology, Hat Yai, Songkla
The young southern university is joining the Botanical Institute in the Zingiberaceae-Project. With some 200 species, Thailand has probably the worlds richest zingiberaceous flora with 4/5 of the genera and 1/6 of all species.

P. Universidad Católica del Ecuador (PUCE), Quito
The close collaboration with PUCE is the cornerstone in the AAU-Ecuador project. The first contact between institutions was established in 1973. In 1979, an extensive teaching and herbarium development program was initiated. A very important element of mutual benefit is the exchange of students and young professional botanists.

Museo Antropológico del Banco Central, Guayaquil

Museo Antropológico has been an inspiring and most valuable counterpart in ethnobotanical surveys of indigenous groups in the northwestern and eastern lowlands of Ecuador. Further projects are being developed.

Ministerio de Agricultura y Ganadería (MAG)., Oficina de Parques Nacionales y Vida Silvestre., Quito and Loja

Knowledge of the biological resources of National Parks and Ecological Reserves is a prerequisite for sound management and planning. The collaboration with staff members of the office of National Parks and Wildlife has been most valuable for the development of various projects of the staff and students of the Botanical Institute.

Museo Ecuatoriano de Ciencias Naturales (MECN)/Casa de la Cultura, Quito

Since the establishment of this institution in 1979, joint collection programmes have been undertaken as a contribution to the establishment of national collections.

Central Ecuatoriana de Servicios Agrícolas (CESA)/ Instituto Nacional de Energía (INE), Quito

A joint project of screening and testing local woody species for reforestation and fuelwood supply in rural Andean communities has been carried out in collaboration with these institutions.

Universidad de Loja, Departamento de Botánica (LOJA), Loja

An inventory of the vegetation and floristics of "Parque Nacional Podocarpus" is being undertaken in cooperation with several botanists from this regional herbarium.

Charles Darwin Research Station (CDRS) Galápagos

Staff members from AAU have twice worked as resident botanists at this renown research center where they received much help and encouragement.

Fundación Natura, Quito

Fundación Natura is an important advocate of the ideas of Nature Conservation in Ecuador. On several occasions AAU has worked with Fundación Natura on botanically related issues concerning nature conservation.

CONTENTS: p.

LECTURE

An evolutionary scenario for the genus *Heliconia*

L. Andersson

The genus *Heliconia* is supposed to have arisen from a strelitzioid stock, which was distinct before early Paleocene. The rise of the genus was most likely closely associated with the rise of the hummingbirds and probably took place between the upper Eocene and lower Miocene. It is proposed that the rise and diversification of *Heliconia* was triggered by the establishment of rain forest vegetation of a modern aspect at about that time. Early diversification probably took place in north western South America or southern Central America, where the genus still has its center of diversity. The strelitzioid ancestor was able to respond to the availability of new habitats because it already had a primitive vertebrate pollination syndrome that could be easily readapted to a new kind of more specialized and highly specific pollinator. There is no strong evidence supporting the view that Pleistocene climatic fluctuations had more than marginal effect on the diversity patterns of the genus. Modern species may well have arisen in the Pliocene or even upper Miocene. (*L. Andersson, Department of Systematic Botany, University of Göteborg, Carl Skottsbergs Gata 22, S-413 19 Göteborg, Sweden*)

LECTURE

Patterns of tree species richness in tropical forests

P. S. Ashton

Patterns of variation in plant community species richness have been correlated with a variety of environmental gradients, as well as to historical biogeography, but causes of this variation remain unknown. The case is made that long term predictability and short term equability of the rainfall regime, in the presence of influences which maximize the diversity of conditions within the soil surface and the subcanopy of a forest community, create the conditions most favorable for species richness among terrestrial plants within a biogeographical region. The significance of this for epiphyte communities will be discussed. (*P. S. Ashton, The Arnold Arbotetum of Harvard University, 22 Divinity Avenue, Cambridge, Massachusetts 02138, USA*)

LECTURE

Diversity of the Ecuadorean forests east of the Andes

H. Balslev* and *S. S. Renner

The eastern Ecuadorean lowlands below 600 m elevation consist of a plain of ca. 130,000 km^2 on the western edge of the Amazon basin. It is covered by tropical rainforest with a rainfall regime of 2,000 to 6,000 millimeters per year and without a pronounced dry season. A total count based on herbarium collections in AAU, F, GB, MO, NY, QCA, S, and US has resulted in a number of 2,700 species of angiosperms. Additional collecting may bring this number up to 3,000. The most important families are Rubiaceae (166 species), legumes (162 species), Melastomataceae (154 species), Arecaceae (80 species), Moraceae (79 species), Lauraceae (76 species), Solanaceae (62 species), Annonaceae (62 species), Gesneriaceae (61 species), Araceae (61 species), Myrtaceae (59 species), and Piperaceae (58 species). In a sample of 1,224 trees above 10 cm dbh, 333 species in 55 families were encountered. Of 536 Ecuadorean species treated in Flora Neotropica and other monographs, 24 percent occur in the eastern lowlands. Using this figure and our total count for the eastern lowlands, we calculate 11,250 to 12,500 species for all of Ecuador, which is considerably lower than previous estimates. (*H. Balslev and S. S. Renner, Botanical Institute, University of Aarhus, Nordlandsvej 68, DK-8240 Risskov, Denmark*)

POSTER

Subfamily Phytelephantoideae (Arecaceae)

A. Barfod

The poster presents the results of a doctoral thesis project initiated in 1985. Four genera and eight species make up the palm subfamily Phytelephantoideae. They are distributed from Panama to Peru reaching western Venezuela and Brazil in the state of Acre. The ecological amplitude of phytelephantoid palms is wide. *Phytelephas macrocarpa* occurs in humid, periodically inundated varzéa forest in the upper Amazon basin, whereas populations of *Phytelephas Schottii* grow in hot areas in the lower parts of the Río Magdalena Valley in Colombia where seasonal drought prevails. The subfamily Phytelephantoideae has been poorly studied since the first species were described from Peru in 1798 by Ruíz and Pavón. This has resulted in overdescription and confusion regarding the identity of the many names published. Phytelephantoid palms are unusual in many ways compared to other groups of palms and they probably represent a separate evolutionary line within the palm family. The dioecious plants have strongly dimorphic flowers. The perianth of the male flower is reduced to merely a rim, whereas the length of the female perianth may reach 25 cm. The hard endosperm of the seed is the source of ivory nut. From the middle of the 19th century until the outbreak of World War Two, large quantities of this material were exported from South America to the United States, Italy and Germany. Ivory nut was mainly used to make buttons. Today it has been replaced by plastic and its importance is limited to cottage industries in the former exporting countries producing mostly souvenirs *(A. Barfod, Botanical Institute, University of Aarhus, Nordlandsvej 68, DK-8240 Risskov, Denmark)*

LECTURE

Architectural concepts for tropical trees

D. Barthelemy* and *F. Hallé

The architecture of a plant depends on the nature and on the relative arrangement of each of its parts; at any given time, it is the expression of an equilibrium between growth processes and exogenous environmental constraints. Considering the plant as a whole, from its germination to its death, the architectural analysis is essentially a global and dynamic approach to plant development. The concepts of architectural model, architectural unit, and reiteration that will be discusssed here, have led to a general interpretation of the growth and architecture of tropical trees and these three major concepts appear to be of particular interest in the understanding of crown construction in trees however complex they are. *(D. Barthelemy and F. Hallé, Laboratoire de Botanique, Institute Botanique, 163, Rue A. Broussonet, 34000 Montpellier, France)*

POSTER

Species richness of the Colombian palm flora

R. G. Bernal

Continued refinements in the count of Colombian palms give a total of 48 genera and 266 species. Species lumping in the last few years has been balanced by new records or the discovery of new species; thus the above figures will most likely not change dramatically with further research. Extensive exploration of Amazonia, however, could result in a small increase. The largest genera are *Geonoma* (48 spp.), *Bactris* (40), *Aiphanes* (22), *Chamaedorea* (16), *Catoblastus* (14), and *Prestoea* (10). Recent extensive exploration has made the Chocó region the best known in Colombia in terms of palms, and has greatly improved our knowledge of the palm floras of the Central and Eastern cordilleras and of the Río Magdalena valley. The eastern Plains, the Amazon region, and the Sierra Nevada de Santa Marta remain the most poorly explored areas. Current research has improved our knowledge of such large and poorly known genera as *Aiphanes*, *Astrocaryum*, *Catoblastus*, *Ceroxylon*, and *Prestoea*. Further research is needed in large genera like *Bactris*, *Chamaedorea*, and *Geonoma*, and in several of the smaller ones.(*R. G. Bernal, Instituto de Ciencias Naturales, Universidad Nacional de Colombia, Apartado 7495, Bogotá, Colombia*)

POSTER

Conservation of *Attalea colenda*, a threatened high yielding oil palm from western Ecuador

U. Blicher

Attalea colenda is a high yielding, but little known oil palm from northwestern South America. It is distributed from the Department of Nariño in southwestern Colombia through the coastal plain of Ecuador below 600 m to the Peruvian border. The palm grows in areas with a yearly precipitation of 500-2,400 mm in very dry forest, dry forest and moist forest. Individuals are often left over when forest is cleared. It was described in 1942 as *Ynesa colenda* by Cook, who recommended it as a potential new source of oil. It is never cultivated, but natives have used the fruits for oil extraction. The palm was largely forgotten until 1987 when its correct taxonomic status was clarified.

A population study of *Attalea colenda* was initiated in March 1987. Two hectare plots, one in natural forest and one in a pasture, were established in northwestern Ecuador. The studies showed that *Attalea colenda* depends on mature forests for regeneration. Adult individuals produce 1-4(-6) infructescences per season, each with 3,500-7,700 fruits. The kernels of *Attalea colenda* contain 53 % fat with a distribution of fatty acids similar to that from the African Oil Palm. One hectare with 130 trees can produce an estimated 1.2 - 10.4 metric tons of oil per year, which is close to production yields from the African Oil Palm.

Attalea colenda populations depend on closed forest to regenerate and the large scale destruction of natural forest is a threat to this potential vegetable oil resource.

(U. Blicher, Botanical Institute, University of Aarhus, Nordlandsvej 68, DK-8240, Risskov, Denmark)

POSTER

The genus *Aiphanes* (Arecaceae) in Ecuador

F. Borchsenius

This study was initiated in January 1987 with the aim to resolve the taxonomy of the genus *Aiphanes* (Bactridinae) in Ecuador. Members of *Aiphanes* are small to moderately sized, spiny palms, with praemorse pinnae. The diverse genus includes a total of some 45 species, distributed from the West Indies and Panama to Bolivia. Twelve species occur in Ecuador, four of which are new to science. A taxonomic treatment of *Aiphanes* in Ecuador will be published in the near future. (*F. Borchsenius, Botanical Institute, University of Aarhus, Nordlandsvej 68, DK-8240 Risskov, Denmark*)

POSTER

The genus *Muehlenbeckia* in South America

J. Brandbyge

The Pacific subantarctic genus *Muehlenbeckia* comprises about 25 species ranging from New Guinea through Australia and New Zealand to temperate South America.

A taxonomic revision of the genus in South America has resulted in the recognition of nine species and four varieties. The genus is characterized by polygamo-dioecious flowers and baccate fruits. The ruminate endosperm, however, is the only constant character to distinguish it from *Polygonum* sensu lato. *Muehlenbeckia* seems most closely related to the Asian section *Pleuropterus* .

Pollen of *Muehlenbeckia* was examined by light and Scanning Electron Microscopy (SEM). The South American taxa are rather homogenous with regard to ektexine surface. The pollen grains are isopolar, radiosymmetric, tricolporate, prolate to subprolate, 18-25 x 18-25 µm with punctate-striate ektexines. A group of Australian species around *M. cunninghamii* are distinct from the rest of the genus by having sparsely spinulose and punctate ektexines. The assumption that these morphologically similar, predominantly stem-assimilating species form a natural group, gains support from the palynological data.

Muehlenbeckia in South America is cytologically uniform. The somatic chromosome number 2n = 20 was obtained by counting root tip cells.

All taxa show great morphological variation, and the existence of intermediary forms indicates that hybridization is a common phenomenon in the group.

The southernmost taxon on the continent, *M. hastulata* var. *rotundifolia* is very closely related to the south Australian complex around *M. adpressa*. *(J. Brandbyge, Botanical Institute, University of Aarhus, Nordlandsvej 68, DK-8240 Risskov, Denmark)*

LECTURE

Patterns of dynamics and diversity in peatswamp and kerangas forest

E. F. Bruenig, Huang, Y. -N., Lee H. S. **and** *Ngui S. K.*

Combined ordination by physiognomic and structural features of 47 individual small stands on different soils and sites, 40 spread over Sarawak and Brunei and seven in south China, produced distribution patterns which were related to soil and site gradients. Patterns of species and physiognomic feature distribution within larger stands (Sabal F.R. 20 ha, 10 cm min. girth, Dalam F.R. 62 stands/plots, 5 cm min. girth) on typically heterogenous soils show complex patterns related to gap sizes and soil variation. Present states of regeneration can be misleading if used for assessing long-term stand dynamics as a result of a patterns of periodicity of gap formation and the effects of sporadic episodic events. *(E. F. Bruenig and Huang, Y.-N., University of Hamburg Leuschnerstr. 91, D 2050 Hamburg 80 FRG; H. S. Lee, Deputy Director Research Forest Department Kuching, Sarawak, Malaysia; S. K. Ngui Amenities and Conservation Branch, Forest Department, Kuching, Sarawak, Malaysia)*

LECTURE

Spanish floristic exploration in America: past and present

S. Castroviejo

For obvious historic and cultural reasons, Spanish botany has been closely linked to American botany. As early as 1535, Gonzalo Fernández de Oviedo published his "General and Natural History of the Indies", in 1580, Nicolás Monardes published the "Medicinal History of the things brought from our West Indies which are of use in medicine", and in 1615, the work "Four books of the nature and virtues of plants and animals..." was published, which Francisco Hernández had written about 1577.

But for modern botany, the main expeditions are those of the 18th and 19th century, in which abundant plant material was collected and important works were published describing the plants of the territories explored. In chronologic order, the main ones are:

1754-1756, expedition to the Orinoco, headed by José de Ituriaga, with Pehr Löfling.

1777-1778, expedition to the Viceroyalty of Perú, under Hipólito Ruíz, with José Pavón, the Frenchman Joseph Dombey and Juan Tafalla. Results: "Flora Peruviana et Chilensis", "Flora Huayaquilensis" and a very important herbarium.

1782-1808, botanical expedition to New Grenada (Colombia) directed by José Celestino Mutis. Results: "Flora of the Royal Botanical Expedition to New Grenada", an exceptional collection of botanical plates and a very valuable herbarium.
1787-1803, Royal Botanical Expedition to New Spain, under Martín Sessé, with José Vicente Cervantes, José Mariano Mociño and others. Result: "Flora of New Spain", the famous botanical iconography the "Mociño plates" and an important herbarium.
Two expeditions along the Pacific must be added to this summary, both of which collected herbaria of undoubted importance. Today, Spain is increasing its efforts towards the study of the American flora, mainly through the Royal Botanic Garden in Madrid. Since 1977, several collecting expeditions have been organized with a certain heterogeneity of means and objectives, but all producing works and materials from several Hispano-American countries. With increasing clarity, the Spanish efforts are centered in Colombia, Bolivia and Paraguay where several projects are now under way, both autonomously and in collaboration with local institutions. (*S. Castroviejo, Real Jardin Botánico, Plaza de Murillo 2, 28014 Madrid Spain*)

POSTER

Breeding biology of some Thai *Syzygium* species

P. Chantaranothai* and *J. Parnell

As a part of a revision of the genera *Eugenia* and *Syzygium* in Thailand we conducted studies on the breeding biology of *Syzygium samarangens*, *S. jambos*, *S. macrocarpum*, and *S. formosum* in Khon Kaen and Chiangmai provinces during the period November 1987 to March 1988. A number of floral visitors were observed including sunbirds, ants, butterflies and honey bees. Experimental treatments of the *Syzygium* species included full emasculation, bagging, autogamous, geitonogamous, allogamous and open pollination. In total, six treatments were applied and at least 40, more often 100 flowers were pollinated per treatment. The results indicate that all of the species are self-compatible and that the first two cultivated species are apomictic as well. Studies of the floral biology of *S. jambos* and *S. macrocarpum* revealed that they both show similar daily timing of anthesis. The growth pattern of their flowers is also similar and followed a simple sigmoid curve. However the period from floral initiation to the onset of bloom in *S. jambos* was shorter than in *S. macrocarpum*. (*P. Chantaranothai and J. Parnell, School of Botany, Trinity College, Dublin 2, Ireland*)

LECTURE

Zingiberaceae as a model for evolutionary patterns of the herbs in southeast Asian forests

Chen Zhong-yi

The Zingiberaceae has a pantropical distribution. The greatest concentration of genera and species is found in southeast Asia. Most species are found in the undergrowth of the forest. Evolutionary patterns in Zingiberaceae are discussed in this paper.
In Zingiberaceae each genus has its own particular basic chromosome number which reflects different levels of evolution and polyploidy. Vegetative reproduction has also played an important role in the speciation of Zingiberaceae. Cytological and palynological data show that Zingiberaceae may be regarded as a natural group. In this family, Costoideae and Zingiberoideae may have been derived from the same ancestor. Each tribe has its own independent line of evolution. It is proposed that Hedychieae is at a relatively active evolutionary stage. Globbeae is an apomictic complex. Zingiberae and Alpineae seem to have a diploid and a tetraploid level in evolutionary trends, respectively. This indicates that Zingiberaceae is still evolving in Southeast Asia. (*Chen Zhong-yi, South China Institute of Botany, Academia Sinica, Guangzhou, Wushan, Peoples Republic of China*)

LECTURE

Speciation patterns in the palms of Madagascar

J. Dransfield

It is well known that the palm flora of Madagascar is particularly rich and interesting when compared with that of adjacent Africa. Although an enumeration of the palm flora has not been completed, recent fieldwork has indicated the presence of new genera and many new species, further emphasizing the diversity of the palm flora. Most of the palms are confined to the humid rain forests of the east coast. Four of the six palm subfamilies have representatives on the island, i.e. Coryphoideae (4 genera, 5 species), Ceroxyloideae (2 genera, ca. 14 species), Arecoideae (15 genera, ca. 113 species) and Calamoideae, only represented by *Raphia*, a possible introduction. Within and between the tribes, numbers of species are very uneven, some tribes or subtribes being represented by few morphologically very isolated taxa in contrast to the arecoid subtribe Dypsidinae which accounts for about three quarters of the total number of species. Within Dypsidinae there is a range of forms from tall tree palms to minute undergrowth palmlets, among the smallest of all palms. Genera have traditionally been separated on number of stamens, shape of anthers and nature of the endosperm, but the use of such characters seems to separate manifestly closely allied taxa. The diversity of the Dypsidinae suggests differentiation of species within Madagascar with little extinction providing the disjunctions which allow the separation of genera. (*J. Dransfield, Herbarium, Royal Botanic Gardens, Kew, Richmond, Surrey, TW9 3AB, England, U.K.*)

POSTER

Epiphytism in lowland rain forest of French Guiana: composition and vertical distribution

R. C. Ek

Within the framework of the "Flora of the Guianas" project an investigation was carried out on the diversity, vertical distribution and ecology of epiphytes in a lowland evergreen rain forest near Saül, in the interior of French Guiana. Thirty trees were climbed and sampled. From trunk base to crown, six height zones were recognized.

The vascular epiphytic flora sampled consists of 175 species in 16 families of seed plants and seven ferns and allies. The most important groups are Orchidaceae with 64 species (36.6 %), Pteridophyta with 33 species (18.9 %), Araceae with 24 species (13.7 %), Bromeliaceae with 19 species (10.9 %), Piperaceae with five species (2.9 %), Bignoniaceae with five species (2.9 %), and Clusiaceae with five species (2.9 %).

Going up from the trunk to the outmost canopy, the number of Araceae decreases whereas Orchidaceae increases in numbers. Diversity and abundance of epiphytes is greatest in lower and middle canopy. The woody hemiephiphytes (e.g., *Oreopanax capitatus*, *Clusia* spp., and *Schlegelia* spp.) also show a preference for the middle canopy and were mostly found rooting in forks and crotches with humus.

The species of vascular epiphytes show the following groups, concerning the preference for height zones: 1) some species behave like specialists, with a striking preference for a particular zone or zones on the tree; 2) Other species have a broader vertical distribution, although often with optima of distribution in certain height zones; 3) A few species appear in (almost) every height zone. (*R. C. Ek, University of Utrecht, Heidelberglaan 2, P.O. Box 80.102, 3508 TC Utrecht, The Netherlands*)

LECTURE

Diversity and distribution of the Guianan species of *Passiflora*

C. Feuillet

The diversity of *Passiflora* species in the Guianas is compared to the other South American regions and these data are related to the distribution patterns of the subgenera. The distribution in America of the Guianan species is illustrated, showing the high endemism in the Guianan region. Then, the emphasis is laid on French Guiana for a study of the distribution patterns of *Passiflora* species in this country, and the unusual species density of *Passiflora* in some localities. (*C. Feuillet, Centre ORSTOM de Cayenne, B.P. 165, 97323 Cayenne Cedex French Guiana*)

POSTER and DEMONSTRATION

Computerized plant cartography in the Neotropics

R. Garilleti, J. Fernández Casas* and *R. Gamarra

A software package is presented, capable of constructing cartographic representations from coordinates contained in a database or in a text file. Originally developed for the Iberian Peninsula, it has now been expanded for use in the Neotropics. The package is able to recover data from any database containing the appropriate field, and from almost any text file or ASCII file with the right structure (in columns). Data can be read both directly in U.T.M. coordinates and in geographic coordinates, which it transforms into U.T.M. It verifies and allows screening of the coordinates, detecting any misspellings, non-existent combinations and those outside the relevant area. On a computerized map of the relevant area, it represents the data on a U.T.M. grid 10 km square, although the package reaches an internal resolution of 100 m square. The final output is a map which may be viewed on screen or printed by plotter or matrix printer.

Computerized maps of Bolivia and Paraguay have been developed. All the distribution maps of the taxa published in the "Flora de Paraguay" series (Genéve, Missouri) have been prepared with this software. (*R. Garilleti, J. Fernández Casas and R. Gamarra, Real Jardin Botánico, Plaza de Murillo 2, 28014 Madrid Spain*)

LECTURE

On the speciation of Malesian Papilionoideae (Leguminosae)

R. Geesink

Immediately after the rediscovery of Mendel's laws, attempts were made to frame the typological speciesconcept and the then developed generalizations from genetics and ecology within Darwin's Evolutionary Theory. We are left with the results: a number of different species concepts, each emphasizing one or a few aspects of species. Attempts to combine the known aspects of species into one "synthetic" species concept seem to obey Gödel's theorem: the logically correct syntheses are incomplete. Perhaps the intuitive desire for an "omnnispective" species concept is unattainable.
A "plurispective" species concept, consistent with evolutionary theory (as the only available scientific theory explaining the diversity of living nature) may be a more satisfactory one, but one cannot avoid to choose among relevant aspects. The ongoing debate as to whether natural species are to be regarded as sets (as structurally defined classes of objects) or as unique, cohesive individuals has, at least, clarified certain problems. The attempts towards a unified theory, recently developed by Brooks and Wiley (their major message is that the evolutionary process must be an entropic process) provides generally applicable fundamental concepts from mathematical/physicochemical sources. Two cases of disjunct species and two cases of sympatric and parapatric speciation (mainly from SE Asiatic Papilionoideae) will be demonstrated in terms of cohesive individuals originated from entropic diversification and entropic decay of their formerly cohesive ancestral species. (*R. Geesink, Rijksherbarium, Rapenburg 70 74, 2311 EZ Leiden, The Netherlands*)

LECTURE

Justicia and *Rungia* (Acanthaceae) in the Indo-Chinese peninsula

B. Hansen

Selected aspects of the occurrence of *Justicia* and *Rungia* in the Indo-Chinese Peninsula, i.e. Cambodia, Laos, Vietnam, Thailand, and Malay Peninsula, are given. With the flora treatments of Clarke (1908) and Ridley (1923) for the Malay Peninsula, of Benoist (1935) for Indo-China, of Imlay (1938) for Thailand in addition to Bremekamp's efforts and especially with Thai Acanthaceae, the total number of species by 1970 were 67 for *Justicia* and eight for *Rungia*. A close

examination of all these species and comparisons with a great number of species from neighboring areas has changed the numbers to 49 *Justicias* and 18 *Rungias*, eight *Justicias* being transferred to *Rungia*, while seven belong in other genera. The rather high number of endemics will be commented on. A more detailed treatment will be given of selected species of *Justicia* from the following sections: 1. sect. Rostellaria: *J. diffusa*, *J. procumbens*, *J. quinqueangularis*. - 2. sect. Rhaphidospora: *J. kampotiana*, *J. scandens*. - 3. sect. Harnieria: *J. neesiana*, *J. quadrifaria*. - 4. sect. Grossa: *J. bicalcarata*, *J. grossa*, *J. prominens*. - 5. sect. Gendarussa: *J. amherstia*, *J. gendarussa*, *J. modesta*, *J. subcoriacea*, *J. ventricosa*.
It is shown that rheophytes occur in at least three of the sections mentioned. (*B. Hansen, Botanical Museum, Gothersgade 130, DK 1123, København K., Denmark*)

LECTURE

Gap-phase dynamics and tropical tree species richness

G. S. Hartshorn

Openings or gaps in the forest canopy are important not only to the structural dynamics of tropical forests, but also to the maintenance of the floristic richness and diversity of many forests. Gap-phase dynamics may be the most important mechanism facilitating the stochastic replacement of fallen trees, the persistance of many tree species within the same ecological guild, and the maintenance of high tree species diversity in many tropical forests. The internal heterogeneity of gaps and subtle edge effects further enhance the richness of regeneration sites due to gaps in the forest canopy.
The role of gaps in tropical forest dynamics provides a model for natural forest management that we are implementing in the Peruvian Amazon. Narrow, 30-40 m strip clear-cuts are rotated through a production forest in such a way that each strip is bordered by intact forest for a majority of the 30-40 year rotation cycle. The close proximity of undisturbed forest and the narrowness of a strip promote excellent natural regeneration of the strip clear-cuts. Results from complete inventories of tree species regeneration on two demonstration strips cut in 1985 will be presented.
The strip clear-cut model for tropical forest management based on gap-phase regeneration not only produces attractive economic benefits to the forest owners, but offers the possibility of using tropical forest without destroying them. Natural

forest management using the strip clear-cut model could be an economically viable and ecologically feasible way of promoting sustainable development of tropical production forest, while contributing to the conservation of biological diversity. *(G. S. Hartshorn, Tropical Science Center, Apartado 8-3870, San José Costa Rica)*

LECTURE

Diversity of Alismatidae in the Neotropics

R. R. Haynes* and *L. B. Holm-Nielsen

The Alismatidae is a subclass of primitive monocots that occurs predominantly in aquatic or marsh habitats, including those habitats in várzea and igapó forests. It consists of three orders and nine families. The subclass is represented in the Neotropics by all nine families and 22 genera. Several genera, for example, *Hydrocleys, Limnocharis, Egeria,* and *Echinodorus* have their centers of distribution and possibly their centers of origin in South America. Similar to aquatic vascular plants in other parts of the world, many species in the Neotropics apparently have wide ecological tolerances and, consequently, are distributed over large areas. Distribution patterns of Alismatidae in the Neotropics include (1) north temperate species that extend into the tropics, (2) species with centers in Central America and the Caribbean Islands, (3) species with centers in northern South America, (4) species with centers in southern South America, (5) south temperate species that extend into the tropics, (6) species restricted to the Andes, and (7) wide ranging species. These patterns are often helpful in interpreting phylogeny. For example, *Hydrocleys* possibly has its center of origin in seasonal dry forests of northeastern Brazil and radiated in these habitats to the south and southwest and then north along the Pacific to northern South America and Central America. *Echinodorus* section *Longipetali* is unusual for aquatic vascular plants in that the species occur in deep shade of the forest floor. The section possibly originated in forests of southeastern Brazil and radiated north and west, resulting in one species in northeastern South America, one species in southeastern South America, and two in the northwestern part of the continent. These routes of distribution illustrate that the taxa probably originated hundreds of thousands of years ago and have had adequate time for speciation and distribution throughout the Neotropics, resulting in a species diversity potentially higher than at the present time. *(R. R. Haynes, Department of Biology, University of Alabama, Tuscaloosa, 35486 Alabama, USA; L. B. Holm-Nielsen, Botanical Institute, University of Aarhus, Nordlandsvej 68, DK-8240 Risskov, Denmark)*

LECTURE

Scrub and low forests in the Venezuelan Guayana, new results on their flora and vegetation

O. Huber

The mainly montane and largely forest-covered Guiana region in southern Venezuela harbours a surprisingly diversified multitude of scrub and low forest vegetation types. In each of the three main altitudinal levels, the macrothermic (lowland), mesothermic (upland) and submicrothermic (summit) level, scrub communities, highly autochthonous floristically and physiognomically, occupy narrow habitat ranges often side by side with other woody or herbaceous formations. Scrubs vary in height between 0.5 and 6 m and the low forests between 5 and 12 m, and both formations present a wide range of crown densities according to substrate and other environmental characteristics. Dominant plant families of the Guianan scrub types are Combretaceae, Theaceae, Euphorbiaceae, and Sapotaceae in the lowlands, Ochnaceae, Theaceae, Vochysiaceae, Burseraceae in the uplands, and Theaceae, Ochnaceae, Asteraceae, and Ericaceae on the summits of the table mountains ("tepuis"). Generic and specific endemicity increases with altitude. Low forests also occur at all altitudinal levels often contiguous to scrub, but do not necessarily represent successional stages of the latter. Certain low forest types occurring in the macro- and mesothermic belts signal the presence of very peculiar edaphic conditions as a consequence of extremely low nutrient levels originating from Guiana parent rock materials such as granites and sandstones. *(O. Huber, C.V.G. and Instituto Venezolano de Investigaciones Científicas (I.V.I.C.), Apartado 80405, Caracas 1080 - A, Venezuela)*

LECTURE

Quaternary geological history of the Amazon

G. Irion

The Quaternary history of the Amazon lowlands is characterized by the deposition of sediments of Andean provenance and by the influence of changing sea-level stages. Most Andean sediments were deposited in the sub-Andean region, but at least one billion metric tonnes per year reached the sea. During the first stages of Pleistocene warm periods, going along with rising sea-levels and damming up of the Amazon drainage system, the deposition of sediments was favoured. The high sea-level heights affected areas as far as 3,000 km upstream the Amazon.

Floodplains corresponding to the different Pleistocene sea-level heights were formed. During low sea-level stages erosion in the drainage areas increased and the water levels of the central Amazon river systems were lowered. Due to the damming-up effect, rising sea-level drowned the valleys and formed lakes in the lower reaches of the rivers and creeks. These lakes remained in river valleys having a low sediment load. Results of ^{14}C-dating of sediment-cores recovered from these lakes correspond closely with the Holocene sea-level curve, when taking the inclination of the water table between the inland lakes and the sea into account.

Areas well above the present water tables have not been reached by Pleistocene high water stages. These areas are intensively weathered since the Tertiary, forming the characteristic "Belterra-formation".

The results of our studies concerning Quaternary geology of the Amazon do not show any climatic change or change in the vegetation cover in the lowlands during the Pleistocene. (*G. Irion, Forschungsinstitut Senckenberg, Abteilung für Meeresgeologie und Meeresbiologie, Schleussenstr. 39a, D-2940 Wilhelmshaven FRG*)

LECTURE

Speciation in the tropics - a case study in the *Asplenium unilaterale* complex

K. Iwatsuki

Biosystematic studies of the *Asplenium unilaterale* complex, or *Asplenium* section *Hymenasplenium*, by various techniques will be taken up as an example of analytical studies of speciation in the tropics.

The fern genus *Asplenium* is also represented in temperate regions and is very suitable for biosystematic studies. However, because the genus is centered in the tropics, which makes information less available, systematic relationships among the species throughout the genus are difficult to clarify. This is the reason why *Asplenium* has remained one of the least known genera among the ferns.

Several aspects of the *Asplenium unilaterale* complex has recently been studied: anatomy, reproductive patterns and morphological variation, biochemical features, cytological features, ecology and morphological variation, genetic diversity, and taxonomy.

Adaptation to wet habitats is evident in this species complex. *Asplenium obliquissimum* is growing under constant spray and this subaquatic form is peculiar in having a bistratose lamina structure without any stomata. Another particular case is found on Seram Island, East Indonesia, where an aquatic form of *A. unilaterale* is found. This aquatic form bears many gemmae.

Asplenium generally has x = 36 chromosomes, but in sect. *Hymenasplenium* chromosome numbers are 38 and 39 or their multiples. Apogamous forms are usually distinct from normally sexual forms. The origin of apogamous forms may be suggested by an analysis using isozymes as a genetic marker.
A screening of free amino acids in *Hymenasplenium* has shown both l- and d-isomers of some amino acids in these plants. It is difficult to trace the metabolic process of the amino acids, though it would be expected to produce fruitful results.
The circumscription of *Hymenasplenium* is another topic to be analysed. Neotropical *A. obtusifolium* and its allies seem to be closely related to *A. unilaterale*. Comparable to *Hymenasplenium* is *A. cardiophyllum* and *Boniniella*. The important characteristics of these species will be compared with those of *Hymenasplenium*. (*K. Iwatsuki, Botanical Gardens, University of Tokyo, Hakusan, Tokyo 112, Japan*)

LECTURE

Flood tolerance and tree distribution in central Amazonian floodplains

W. J. Junk

The monomodal flood pattern of the Amazon river and of its large tributaries results in a single prolonged annual period of high flooding over extensive floodplains along their middle reaches. A flood-level gradient was calculated based on 80 years of waterlevel records at Manaus harbour. Plant distribution along this gradient shows a mean flooding period of ca. 270 days/year for the most flood-tolerant shrub communities. Forest communities begin to grow in areas exposed to flooding for ca. 230 days/year. It is shown that factors other than flood tolerance of adult trees (e.g., seedling establishment, sediment deposition, soil waterlogging, seral stage of the forest community) are also important to explain current tree distribution. Amazonian trees exhibit few morphological adaptations to prolonged, high flooding. Physiological and phenological adaptations are of greater importance. High species diversity in the floodplain forest is explained by the habitat diversity, itself caused and maintained by river dynamics, to the continuous existence of floodplains in Amazonia for geologically long periods of time and to the predictability of the flood pulse in the lower reaches of large rivers. (*W. J. Junk, Max Planck-Institute for Limnology, Working Group for Tropical Ecology, P.O. Box 165, D-2320 Plön, FRG in collaboration with Instituto Nacional de Pesquisas da Amazonia (INPA), 69000 Manaus, AM Brazil*)

POSTER

Structure and composition of a montane forest in Ecuador

P. M. Jørgensen* and *R. V. Reyes

A quadrate of 1 hectare (100 x 100 m) was established between 3,225 and 3,310 m a.s.l. on Mt. Pasochoa, 30 km south of Quito. We found a density of 1,060 trees/ha (DBH ≥ 5 cm) and 710 trees/ha (DBH ≥ 10 cm); 34 species were represented in the forest, belonging to 31 genera and 21 families. The total basal area was calculated to be 25.8 m^2 of which *Miconia theaezans* and *Piper andreanum* account for more than 50 percent. The Family Importance Values (FIV) show that Melastomataceae is by far the most important family followed by Piperaceae and Asteraceae, which only rank so high due to their high diversity. The Importance Value Index (IVI) shows that *Miconia theaezans* is the most important species followed by *Piper andreanum* and *Miconia pustulata*. The most abundant epiphytes are mosses, orchids, bromeliads and ferns: one or more found on 98 % of the trees. Vines and lianas are found on 79 % of the trees. The ground cover is dominated by mosses and *Chusquea scandens*, but many species are codominant. *(P. M. Jørgensen and R. V. Reyes, Herbario QCA, Departamento de Ciencias Biológicas, Pontificia Universidad Católica del Ecuador; Apartado 2184, Quito, Ecuador)*

POSTER

Vegetation dynamics in the western Amazon lowlands: the role of changing flood influences

R. Kalliola, M. Puhakka, M. Rajasilta, M. Räsänen* and *J. Salo

The western Amazon is a fluviodynamic mosaic where large rain forest areas are repeatedly modified by fluvial perturbance. In addition to channel migration and riverine primary succession, a number of other fluvial processes such as floodplain abandoning, river damming and backswamp dynamics occur. They further affect the forest development causing large scale modification and destruction of previous forest generations. Tens of thousands of km^2 of lowlands are under the influences of permanent floods, especially in the Pastaza-Marañon and Ucayali basins. There are two distinct vegetation categories which differ from the closed-canopy tropical forest covering most of the area: 1) aguajales are characterized by the abundance of the palm *Maurita flexuosa* (aguaje), and 2) pantanales, which are treeless swamps. These flood influenced vegetation categories may develop at any forest site close to changing river channels, and they may later disappear as local drainage improves. Major floodplain alterations also modify terra firme areas by causing sudden local flooding. The abundance of open, struggling and even dead forests (riverine liana forests) close to large rivers like the Ucayali suggest that the Amazonian biota is not perfectly adapted to the rate of changes in flooding patterns and drainages. The reconstruction of the western Amazon vegetation history needs more understanding of these types of processes as aggrading rivers have constantly been present in the area. Data on local changes between well-drained and waterlogged types of vegetation may be understood in terms of present floodplain dynamics. Floodplain forest classifications are temporal, because a geographic locality may face site turnover through river erosion and primary succession and later be subject to changing flood influences . (*R. Kalliola, M. Puhakka, M. Rajasilta, M. Räsänen and J. Salo, Department of Biology, University of Turku, SF-20500 Turku, Finland*)

POSTER

Taxonomical aspects of scanning electron microscopic (SEM) studies of leaf surfaces of Cuban *Buxus* species (Buxaceae)

E. Köhler

According to recent SEM investigations the sculpturing types and patterns of plant surfaces are under genetic control and seem to be suitable for taxonomic characters between the supra- and infraspecific leve. The Cuban species show remarkable differences in leaf surface sculpturing which is briefly demonstrated. It will be shown how these characters can contribute to the identification and delimitation of the species and to add to the knowledge of their relationships which is based mainly on pollen morphological, leaf venation and leaf anatomical characters so far. In combination with other features the leaf surface pattern of the Cuban *Buxus* species represents a most valuable help for taxonomy. (*E. Köhler, Ber. Botanik u.- Arboretum, Museum für Naturkunde der Humboldt-Universität, Späthstr. 80/81, DDR-1195 Berlin, GDR*)

POSTER

Phytosocoiological methods - results of two analyses in a species rich Amazonian rain forest

J. Korning

This inventory is part of the AAU Ecuador-project concerning vegetation studies in tropical rain forest. Phytosociological results for trees ≥ 10 cm DBH obtained in a quadrate plot are compared to published results obtained by the Point Centered Quarter Method along a line transect at the same locality in Añangu, Amazonian Ecuador. The line transect had 728 individuals ≥ 10 cm DBH, 239 species, 51 families, a total basal area of 35.7 m^2 and an estimated above ground phytovolume of 432.2 m^3 per hectare. The one hectare quadrate plot of 100 x 100 m had 734 individuals ≥ 10 cm DBH, 153 species, 46 families, a total basal area of 22.2 m^2 and an estimated above ground phytovolume of 240.5 m^3. Among the 20 most important species, only four were common to both plots. The most important species were *Quararibea ochrocalyx* on the quadrate plot and *Iriartea deltoidea* on the line transect constituting 26.6 and 13.3 % of the individuals, respectively. Most important families on the quadrate plot are Bombacaceae, Arecaceae, Moraceae, Caesalpiniaceae, and Lauraceae and on the line transect Arecaceae, Moraceae, Meliaceae, Mimosaceae and Caesalpinaceae constituting 40.4 and 35.4 % of the Family Importance Values of the plots, respectively. The Point Centered Quarter Method used along a line transect reflects maximum variation in the area. The quadrate plot reflects the precise structure and composition of the forest within the plot. (*J. Korning, Botanical Institute, University of Aarhus, Nordlandsvej 68, DK-8240 Risskov, Denmark*)

LECTURE

Speciation, diversity, and the role of Ericaceae in Neotropical montane vegetation

J. L. Luteyn

The Ericaceae are ideally suited to the isolated, ecologically diverse, and youthful montane habitats of tropical America. Within wet, montane regions they have successfully filled open, woody vegetation communities, including disturbed habitats, perhaps better than any other group and are therefore one of the most important families ecologically. In the Neotropics the Ericaceae have diversified and speciated to include about 40 genera and more than 1000 species of terrestrial and epiphytic shrubs (to small trees). This lecture will cover four broad topics: 1) the morphological features of the family, genera and species which are the bases for current taxonomic systems; 2) the distribution of the family in LatinAmerica emphasizing areas of especially high diversity and endemism; 3) biological and ecological factors contributing to the success of the family; and 4) the role of the Ericaceae in Neotropical montane vegetation. (*J. L. Luteyn, New York Botanical Garden, Bronx, New York 10458-5126, USA*)

LECTURE

Speciation trends in neotropical Annonaceae

P. J. M. Maas

The Annonaceae are a large and rather homogenous family, distributed throughout the tropics and generally easy to recognize. Most members occur in forests while a minority prefer more open or even xeric habitats. The main altitudinal range is from sea level up to 1000 m, though here, too, there are some notable exceptions. Hardly anything is known yet about factors affecting speciation. Field observations have been made in recent times, e.g., by Gottsberger (pollination) and Morawetz (ecology). All considered, however, systematics necessarily still has to rely upon morphology.

Intergeneric diversity is demonstrated in two of the largest groups of neotropical Annonaceae, namely the *Guatteria*-group (Annonoideae - Uvarieae) and the *Annona*-group (Annonoideae - Unoneae). Intrageneric diversity is especially focussed here on the genera *Guatteria* (over 200 species) and *Rollinia* (ca. 50 species).

In the *Guatteria*-group, distinction between genera is mainly based on petal and fruit characters. For specific delimitation in *Guatteria,* presence or absence of tiny warts on the leaf surface, shape of monocarps, and characters of indument, petals and stamens are significant.

In the *Annona*-group, distinction between genera is based on characters of stamens, petals, and fruit. In *Rollinia* the most important differentiating characters are found in the fruit, the corolla wings, leaf indument, and presence or absence of a spur-like hump on the sepals. (*P. J. M. Maas, University of Utrecht, Heidelberglaan 2, P.O. Box 80.102, 3508 TC Utrecht, The Netherlands*)

POSTER

Plants and man on Puná Island, Ecuador

J. E. Madsen, R. R. Mix* and *H. Balslev

Puná Island (919 km^2, population 10,000) in the delta of the Río Guayas on the west coast of Ecuador was an important site in pre-Columbian South America. The Museo Antropológico in Guayaquil has a long lasting research interest on the island, but lacked baseline information about its natural setting and plant resources available to the cultures which developed there between 5,000 and 500 years ago. During 1987 a combined team from the University of Aarhus and Museo Antropológico collected data about the plant life of the island, including information on uses and plant names among the present mestizo population. The natural vegetation of Puná Island includes dry deciduous forest, semi-evergreen forest, savanna, salt marshes, mangroves, sandy beach vegetation, and aquatic vegetation around ponds and small lakes. A total of 442 species of plants were recorded and collected during the fieldwork. The mestizo population appears to have conserved much of the knowledge about plants possessed by pre-Columbian cultures on Puná Island. The older people on the island know and have names for most of the species encountered. They use many of them for consumption, construction, medicine, weaving and a wealth of other purposes. A small ethnobotanical herbarium was established in the local school to help the teacher educate the children about their natural heritage. *(J. E. Madsen and H. Balslev, Botanical Institute, University of Aarhus, Nordlandsvej 68, DK-8240 Risskov Denmark; R. R. Mix, Museo Antropológico del Banco Central, Avenida 9 de Octubre y José de Anteparra, Casilla 1331, Guayaquil, Ecuador)*

POSTER

Oil company cooperation in biological research in Ecuador

D. McMeekin

There is now a joint endeavor between an oil company and biological scientists working within the Amazon region of Ecuador in developing a broader understanding of the regions tropical forests. This cooperation is an effort to achieve a balance between forest preservation and the wise use of natural resources.

Conoco Ecuador Ltd. is providing helicopter transportation, housing and food to scientists to visit virgin areas accessible only by air within their concession block of 200,000 hectares. Initial investigations have identified species of trees, orchids and amphibians new to science. Oil reserves in Ecuador are going to be developed, a reality that must be recognized and accepted. Utilizing logistical opportunities now available to biologists, gathering and disseminating information on the region and working together toward the best way to develop natural resources will help insure the protection of vast areas of Ecuador's Amazon forest. The example set in Ecuador of cooperation between corporations and biological scientists is providing valuable training for Ecuadorian nationals and research opportunities in all fields of biology. The understanding and information gained from these combined efforts will further the cause of tropical forest preservation. *(D. McMeekin, Conoco DU 3104, P.O. Box 2197, Houston, Texas 77252, USA)*

LECTURE

Diversity of Lecythidaceae in the Guianas

S. A. Mori

A total of 50 species of Lecythidaceae occur in Guyana, Suriname, and French Guiana (The Guianas). Ten species (20%) are endemic to the Guianas and 23 (46%) are endemic to the Guiana floristic province. *Couratari guianensis*, *Couroupita guianensis*, *Eschweilera coriacea* and *Lecythis corrugata* are found in the Guianas as well as west or north of the Andes. In addition, *Gustavia hexapetala* (east), *G. dubia* (west), *Couratari stellata* (east), *C. scottmori* (west), and *Lecythis zabucaja* (east), and *L. ampla* (west) are species pairs which indicate the transAndean relationships of Lecythidaceae. Only the widespread *Gustavia augusta* of the Guianas is found in the coastal forests of eastern extra-Amazonian Brazil. All Lecythidaceae of the Guianas are found in the lowlands. Only *Lecythis alutaceae* has been collected at elevations above 1,000 meters. Most Lecythidaceae of the Guianas inhabit non-flooded (terra firme) forests. However, some species are restricted to areas with moister soils (e.g., *Couratari gloriosa* and *Lecythis pneumatophora*) and others are restricted to savanna habitats (e.g., *L. brancoensis* and *L. schomburgkii*). Species of Lecythidaceae with actinomorphic androecia are poorly represented in the Guianas. The only genus with this andreocial type is *Gustavia* and it is represented by only three species. A study of the 27 species of Lecythidaceae found in the proposed national park surrounding Saül, French Guiana, has shown that closely related species in this area may occur in different habitats, be pollinated by different animals, bloom sequentially, or have their seeds dispersed either by the wind or different animals. A case is made for the preservation of the 133,600 hectare proposed national park surrounding Saül because by so doing 54% of the known diversity of Guianan Lecythidaceae will be protected. (*S. A. Mori, New York Botanical Garden, Bronx, New York 10458-5126, USA*)

POSTER

Agroforestry in Amazonian Ecuador: the contribution of botanical research

D. A. Neill

The human population of lowland Napo province in Amazonian Ecuador has increased more than four-fold in the past 20 years, as a consequence of road construction for petroleum development in the region, and now stands at about 120,000. Most of the new colonists obtained 50-hectare parcels through the Ecuadorian government's land distribution programs. Part or all of the original forest has been cleared from each of these small farms, and the colonists cultivate

commercial crops such as coffee or cacao, subsistence crops such as manioc, and pastures for beef cattle.
The Ecuadorian Ministry of Agriculture and the U.S. Agency for International Development are carrying out an extension agroforestry program for the Napo colonists. Native species of fast-growing timber trees are grown in combination with crops or pasture. The harvest of this timber will provide the farmers with additional income and, it is hoped, will relieve pressure for timber exploitation from the remaining primary forests of the region.
Missouri Botanical Garden, in conjunction with the Napo agroforestry project, is carrying out field research on the trees of the region. This project has several components: 1) Floristic-dendrological -- an inventory of the tree species of the region is being made through intensive collecting of herbarium specimens. Based on this work, a Spanish-language field guide to the trees of the region is being written and will be published in Ecuador. 2) Ecological -- structure and composition of primary forest is being studied through establishment of permanent 1-hectare plots at several localities. Also, studies of reproductive phenology and natural regeneration of desirable native timber species are being carried out. 3) Silvicultural -- experimental arboreta for promising tree species are being established, and growth rates monitored. Also, experimental studies of propagation by seeds or cutting are being done. The research results are helping to provide a biological information base for the Napo agroforestry project, and for other development and conservation projects that may be carried out in the Napo region in the future. (*D. A. Neill, Missouri Botanical Garden, USAID / Quito, AID, Washington DC 20523, USA*)

LECTURE

Eco-units as subsystems of rain forest mosaics

R. A. A. Oldeman

Eco-units are defined as surfaces on which forest development has started at one moment due to some event. This forest development can be analysed according to architectural, ecophysiological and species composition criteria, and is always determined by one set of trees from the beginning to the end. Eco-units are developing on all forest regeneration surfaces, from small gaps to large destroyed terrains, and may be created by many different factors, from gusts of wind to earthquakes and fire. Eco-units will be described as living systems, but not superorganisms. This description includes their border layers, their internal systems of signals and translocation, their adjustment to outer circumstances and their mode of reproduction. Finally, the concept of eco-unit borders will be critically surveyed, comparing a discrete model of interacting eco-units in a forest mosaic with a continuous model involving gradients and interference patterns (*R. A. A. Oldeman, Agricultural University, Silviculture and Forest Ecology, P.O. Box 342, 6700 AN, Wageningen, The Netherlands*)

LECTURE

Speciation characteristics of African Loranthaceae

R. M. Polhill

The patterns of speciation in African Loranthaceae demonstrate general principles important to plant conservation and development of natural resources, as well as to concepts of evolutionary biology. Most species are precisely confined to major chorological divisions that were postulated originally from the tree flora. The distributions of loranthaceous aerial hemiparasites owe little to climate or geomorphology, but much to the birds that pollinate and disperse them. Despite natural and human disturbances, the ecosystems of Africa have high levels of complexity and integrity; forest ecosystems are particularly fragile.

Like other groups in Africa, a few relatively primitive genera show patterns of geographical spread that bear little relation to modern vegetation zones. More modern groups follow two basic patterns - 1) equatorial, spreading into drier habitats north and south; 2) longitudinal along the Rift Valley of eastern Africa, spreading little westwards and showing more xeric adaptations even when in forests now.

Pollination mechanisms in African Loranthaceae reach exceptional levels of specialization for the family and show remarkable parallels in different evolutionary lines. The morphological radiation in Africa is notably different to that of Asian relatives and appears to depend on the conditioning of the avifauna.

(*R. M. Polhill, Herbarium, Royal Botanic Gardens, Kew, Richmond, Surrey TW9 3AB, England, UK*)

POSTER

Forest mosaicism and forest patch dynamics in the western Amazon floodplain

M. Puhakka

Recent studies have shown that the western Amazon rain forests have been modified by fluvial perturbances since the Tertiary. Fluvial dynamics have been suggested to operate as an intermediate disturbance factor which prevents the riparian forests from reaching equilibrium: the number of species is high and the communities are diverse because there is no competitive exclusion. However, the packing of forest patches due to channel cut-offs has achieved lesser interest, although the heterogeneity of the area also adds to the species numbers. In this work, forest patch dynamics in two example areas at the rivers Ucayali and Madre de Dios in Peru are described.

Packing of successional forest patches created by loop cut-offs and other channel

alterations create a forest mosaic with sites of varying ages and origins. In addition to forest development on the depositional bars, floodplain environments also contain patches of forests originated from abandoned river channels (e.g., oxbow lakes and swale lakes). Mixing of these forest units of varying origins diversifies the overall forest structure: more habitats are opened than would be available in the case of an even-aged forest bed. At the river Ucayali the old meander forest within the mosaic forest areas are easily distinguishable on SLAR images because of their large ridge-swale structure. The alternation of low lying swales and forested ridges creates long sharp boundaries (ecotones) between forest and treeless areas. The ridge-swale alternation is not that extreme at the Madre de Dios river, but the same patterns are found there on a smaller scale. The inner heterogeneity of the oxbow lakes further diversifies the structure of mosaic forests through floating successions, secondary channel formations and internal deltaic formations. Flooding and overbank sedimentation may later homogenate the mosaic forest structure by destroying its ridge-swale topography and by creating backswamps. Abandoned floodplains, river channels and backswamps make sharp boundaries with the neighbouring forests at higher terrace levels adding to the overall heterogeneity of the floodplains.
The mosaic nature of the western Amazon floodplain forests has been recognized in many recent works (e.g., studies of birds and monkeys at Cocha Cashu, Peru), however, the fluvial dynamics behind the mosaicism has not yet been fully described or understood. The western Amazonian forests have the most diverse floral and faunal assemblages in the world, but it has not always been realized that this may reflect the small-scale mosaic structure of the forests. Obviously the mosaic nature of the floodplains contributes to the high species richness in the Amazon area by providing sites with specialized vegetation ("between habitat diversity") (*M. Puhakka, Department of Biology, University of Turku, SF-20500, Turku, Finland*)

POSTER

Occurrence of riverine sequential successional forests in the Peruvian Amazon

M. Rajasilta

The channel migration of lowland rivers rich in suspended solids is a basic phenomenon in the western Amazonia. The rivers carry eroded sediments and deposit them along river margins where these deposits are open to plant colonization. The following primary succession results in a zonated forest structure (sequential successional forests) at point bars. The early pioneer communities are dominated by *Tessaria integrifolia* (Asteraceae), *Gynerium saggitatum* (Poaceae) and *Cecropia membranaceae* (Cecropiaceae) and later by *Ficus* spp. (Moraceae) and *Cedrela odorata* (Meliaceae).

As the geomorphic origin is common to all riverine successional forests, the numerous patches of these forests offer an unique opportunity to survey factors of rain forest regeneration and diversification. Most of the western Amazon lowland rivers have high rates of channel migration. Thus, the habitat is repeated along the river courses in the area. It is not yet known how similar the plant communities are, but according to preliminary observations, high similarity is to be expected. The spatial distribution and within-patch dynamics of the successional forests is a question of great interest in biogeography: the archipelago-like pattern of the successional forests allows a survey of factors contributing to the species diversity.
In the present study, the occurrence and dynamics of the sequential successional forest in the Peruvian Amazon were surveyed using different remote sensing methods (SLAR, aerial photographs and Landsat MSS and TM imagery) backed by field observations in 1986-87. Data from these surveys are presented. (*M. Rajasilta, Department of Biology, University of Turku, SF-20500, Turku Finland*)

POSTER

The importance of early drawings and paintings as documents for the rapidly vanishing tropical forest vegetation

C. Riedl-Dorn

In the late 18th and early 19th century it was costumary that artists, landscape painters such as F. Bellermann, Th. Ender, J. Selleny and many others as well as botanical illustrators like F. Bauer and J. Buchberger accompanied scientific expeditions to document botanical, zoological, ethnological, and other kinds of observations. Also, many scientists like Ph. Commerson, C. Ph. V. Martius, E. Poeppig, etc., were gifted artists whose sketches executed in the field served as models for the final representation for publication by specially employed illustrators at home in some instances. Due to bad light and other disadvantages in primeval forests, drawings are often far superior even to photographs which finally took their place. Only a small percentage of such drawings and paintings has ever been published, especially of the original versions prepared in the field with their higher degree of authenticity. Much of the areas covered by nearly impenetrable jungle in the first half of last century or even later are under cultivation or urbanization now. The pictures executed in the course of expeditions which are now in natural history museums, art galleries and private collections, should be made accesible and catalogued for the reconstruction of natural forest vegetation in the tropics as a valuable tool for studies in plant geography and synecology. (*C. Riedl-Dorn, Vienna Natural History Museum, Department of Botany, Burgring 7, A-1014 Vienna, Austria*)

LECTURE

The western Amazon basin: a fluvio-dynamic mosaic

J. Salo

Most of the western and central Amazon lowland rain forests are located on fluvially deposited plains. These cyclic fluvial deposits cover older, more weathered and consolidated strata forming a continuous, 20-40 m thick geomorphologically and stratigraphically uniform surficial alluvial bed. The present terra firme rain forests grow on the dissected surfaces of these Pleistocene deposits. Thermoluminescence dating shows that the terra firme-level has been aggraded during the Pleistocene after the final impulses of Plio-Pleistocene compressional shortening and thrusting along the eastern sub-Andean fold and thrust belt. During the Quaternary, the fluvial plains of western Amazon have faced postdepositional deformation resulting in altered hydrology with regional denudation, drainage reversals, river capturing, relocation and terrace formation.
In the western Amazon lowlands, the following landscape processes operating at various levels of terrestrial and aquatic biotic communities are evident due to the mobile geological setting:
1) River dynamics as a disturbance factor. In the western Amazon, lateral channel erosion affects floodplain biota by destroying the old forest and by initiating primary succession on deposited river sediments. By doing this, the river erosion prevents competitive exclusion within various forest types. However, as these river dynamics have characterized the area since the Tertiary, and because of the relatively predictable nature of channel processes within the present floodplain, perturbance by lateral erosion is not a truly unpredictable disturbance factor in terms of the intermediate disturbance hypothesis.
2) Within-floodplain site-turnover and forest patch packing. The mosaic forest structure characterizing the western Amazon varzéa forests is created by channel diversions within the contemporary flood basins. The packing of differentially aged forests diversifies the overall forest structure because more habitats are opened than would be available in the case of an even-aged forest bed. As the channel diversion processes are more unpredictable than those in the lateral migration of the channel, community structure of mosaic forest is highly complex.
3) Floodplain diversions. The aggradational floodplains in the basins of Pastaza-Marañon and Ucayali show a high degree of contemporary floodplain abandonment and diversion. These activities leave behind a network of abandoned floodplains and they also have been present during the history of the Acre and Madre de Dios basins.
4) The foreland molasse basins and their marginal denudated areas. The current scheme of western Amazonia is composed of four major molasse basins separated by slightly higher areas exhibiting surface erosion (denudation). It is obvious that even this level of mosaicism, with the largest components, represents dynamic

Pleistocene relief evolution due to tectonic or downwarp-induced shift between erosional and depositional surfaces.
These levels of landscape processes and the resulting mosaicism represent a genuine chronological continuum. Biologically the continuum may also represent a continuum from an ecological to an evolutionary theater. (*J. Salo, Department of Biology, University of Turku, SF-20500 Turku, Finland*)

POSTER

Changing flora and vegetation on the summits and ridges of Doi Chiangdao, the massive limestone mountain in northern Thailand

T. Santisuk

Doi Chiangdao, the massive Permian limestone mountain (c. 450 - 2,190 m a.s.l.) in Chiangdao District of Chiangmai, northern Thailand, harbours a distinct flora and represents an astonishing reservoir of life forms not found elsewhere in Thailand. These mountains areas covering some 528 km^2 have been under management of the Division of Wildlife Sanctuaries, Royal Forest Department since 1978. Tropical, subtropical and temperate species are encountered in varying altitudinal zones of Doi Chiangdao. Exposed rocky peaks and ridges carrying subalpine-like vegetation abound with temperate genera and species (e.g., *Cotoneaster*, *Delphinium*, and *Scabiosa*). Some species mark their southernmost distribution in Doi Chiangdao, e.g., *Cotoneaster franchetii* and *Inula rubicaulis*, and certain species are endemic to this mountain, representing typical plants of subalpine and alpine vegetation (e.g., *Saxifraga gemmipara* var. *siamensis* and *Scabiosa siamensis*).
Illegal conversion of the montane forests at elevations above 1,100 m a.s.l. into shifting cultivated fields by nomadic hill people in recent decades has resulted in drastic gully erosion owing to the steepness of this mountain. The annual burning in the dry season has been disastrous to the existence and dispersal of these temperate species on the exposed limestone summits and ridges. This unique vegetation has also been threatened recently by the increasing number of undisciplined hikers. A number of noxious weeds (e.g., *Ageratina adenophora* and *Arundinella setosa*) have developed extensively over the disturbed areas replacing native flora. Complete changes in vegetation type, flora and fauna on the summits and ridges of Doi Chiangdao will be the likely result in the near future, if no immediate and effective action is taken to relieve these biotic pressures. (*T. Santisuk, The Forest Herbarium, Royal Forest Department, Bangkok 10900, Thailand*)

LECTURE

Diversity of minor tropical tree crops and their importance for the industrialized world

V. K. S. Shukla* and *I. Nielsen

The minor tropical tree crops play a major role for the modern industrialized chocolate world. Chocolate is associated with an important commodity, the cocoa bean. This tree is cultivated in West Africa, South America, Central America and the Far East. The uncertainty in cocoa butter supply and the volatility in cocoa butter prices depending on the fluctuating cocoa bean prices have forced confectioners to seek other alternatives which may have a stabilizing influence on the prices of cocoa butter. The continued research in the field of confection science resulted in the development of various new exotic fats such as Sal, Shea, Illipe and Kokum. This lecture describes in detail the use of these exotic tree crops in the chocolate industry for improving the quality and the cost economics. (*V. K. S. Shukla, Analytical Research & Development, Aarhus Oliefabrik A/S, P.O. Box 50, DK-8100, Aarhus C., Denmark; I. Nielsen, Botanical Institute, University of Aarhus, Nordlandsvej 68, DK-8240 Risskov, Denmark*)

LECTURE

Zingiberaceae of Thailand with emphasis of *Boesenbergia* and its allies

P. Sirirugsa

Zingiberaceae comprises approximately 30 genera and 200 species in Thailand. Selected representatives of Thai Zingiberaceous species, *Zingiber spectabile*, *Curcuma domestica*, *Etlingera littoralis*, are presented. The genus *Boesenbergia* of Thailand has been revised and 13 species are recognized. Three of them, *B. acuminata*, *B. basispicata*, and *B. petiolata* are newly discovered. *B. siamensis* has been transferred from *Gastrochilus siamense*. The other eight species of the genus are: *B. angustifolia*, *B. curtisii*, *B. longipes*, *B. parvula*, *B. plicata*, *B. prainiana*, *B. pulcherrima*, and *B. rotunda*.

The morphological characters of the new species and their relatives, as well as the distinguishing characters of some species of the genus, are described. The variation of the species *B. plicata* and *B. rotunda* are discussed. Some species of the genus *Kaempferia*, which is allied to *Boesenbergia*, are also presented and their variation is discussed. (*P. Sirirugsa, Prince of Songkla University, Hat Yai 90110, Thailand*)

POSTER

A revision of *Hyospathe* (Arecaceae)

F. Skov

Hyospathe is a genus of small, spine-less understory palms belonging to subtribe Euterpeinae. Only two species are recognized: *H. macrorhachis* and the very variable *H. elegans*. Members of the genus are found in moist, tropical rain forest from Panama to Bolivia and from coastal Colombia to Pará in northeastern Brazil. *Hyospathe* is particularily common in the foothills along both sides of the Andes, with an altitudinal limit of about 2000 m a.s.l.
Prior to this revision, a total of 20 names were known in the genus, most of which were published early in this century. Due to a narrow species concept prevailing at that time, species were separated on subtle differences in quantitative characters (e.g., length and width of leaves and inflorescences, division of leaves, etc.). Intensive fieldwork in Ecuador and examination of 330 vouchers showed, however, that *Hyospathe* exhibits a very wide and continuous variation. Separation of species using the above mentioned characters proved impossible and most of the names were synonymized under *H. elegans*. This "mass slaughter" of names was also justified by palynological and anatomical evidence.
Morphological diversity in *H. elegans* has been given a complete and thorough treatment, because broadly defined species tend to obscure geographic patterns of specific variation. (*F. Skov, Botanical Institute, University of Aarhus, Nordlandsvej 68, DK-8240 Risskov, Denmark)*

DEMONSTRATION

Hypermedia - a new approach to computerized herbarium taxonomy

F. Skov

In taxonomy, large amounts of data must be stored and organized in order to ensure maximum utilization. Taxonomic data are complex: they include text, numbers, maps, photograps, drawings, etc. These are all connected in a weblike, nonhierarchical way, which makes it difficult for computers to handle them.
Hypermedia is an example of a new software-generation capable of storing and manipulating many kinds of information such as text, graphics, and sounds. Fast information retrieval is one of the basic features: searching can be done without following a predetermined organization scheme; instant branching to related facts is possible. User friendly operation is another characteristic: most actions require no more than pointing and "clicking" on the screen.
This demonstration presents such a Hypermedia application. It is a computer tool designed for the management of all data necessary in taxonomic revisions.

Information is stored on "cards" collected in stacks: One stack contains cards holding information from herbarium specimens, another contains cards with information about names of taxa, and the third comprises a literature index. It is possible to move freely between these stacks and seek information. Additionally, the program creates distribution maps and produces output text-files like reference lists and full-format taxonomic reports.

The program has it limits: 1) it is not as fast as a relational database; 2) it does not possess the calculating power of a spreadsheet; 3) revisions including more than 2-3,000 collections would make the system too slow to be practical. However, its simplicity and user friendliness makes it an efficient tool for reducing much of the time-consuming work connected to data-registration and -handling. Moreover, its fast searching abilities greatly reduce the risk of loosing or accidentally "forgetting" data. *(F. Skov, Botanical Institute, University of Aarhus, Nordlandsvej 68, DK-8240 Risskov, Denmark)*

LECTURE

Dry deciduous forests of the Paraguayan Chaco

R. Spichiger

The Gran Chaco is a vast alluvial plain, covering about 1 million km^2, spread over Bolivia, Paraguay and Argentina. The Paraguayan part of the Chaco represents 247,000 km^2. The climate is characterized by two contrasting seasons: rainy summers and dry winters. A pluviometric and temperature gradient exists from east to west: precipitation decreases from east to west (1,300 mm at the eastern limit and 500 mm at the Bolivian border), while temperatures increase from east to west (mean of 24° C at Asuncion and 26° C in the northwest). The vegetation of the Chaco varies according to the climatic gradient described above:

1) A mosaic of extremely xeromorphic forests, palm savannas, flooded grass savannas and hydrophyllous gallery forests along the streams characterizes the south and southeast. Transects from one of the branches of Río Pilcomayo show, that the layout of vegetation formations differs, depending on the duration of flooding.

2) An extremely xeromorphic forest characterizes the west, the center, the north and the northwest. The following species are particularly numerous in these forests: *Aspidosperma quebrachoblanco*, *Ruprechtia triflora*, *Bulnesia sarmientoi*, *Stetsonia coryne*, *Chorisia insignis*, various *Capparis* and *Acacia*. Due to local circumstances, this forest may develop into types in which only one or two species are predominant.

The following characteristics should furthermore be stressed:

1) The presence of an important northwestern relief (Cerro León, about 800 m high). The vegetation appears more hygrophilic than in the neighboring plain. Transects show a gradient of the vegetation along the slopes (*Piptadenia*

macrocarpa forest).
2) The presence of a fossil hydrographic network, sometimes still partially filled, thus forming isolated lagoons with hygrophilic vegetation and gallery forests. Gramineae are the main inhabitants of the fossil dry sandy beds. They are called "espartillares".
3) The presence of Río Timane at the northwest of the Chaco, whose waters come from the Andean foothills and disappear in the Chaco plain. The gallery forest bordering the river is rich in hygrophilic and riparian species. (*R. Spichiger, Conservatoire et Jardin Botaniques, Case postale 60, CH-1292 Chambésy/GE, Switzerland*)

LECTURE

Population dynamics of tree species in tropical forests

M. D. Swaine

Populations of tree species in tropical forest have been monitored in some studies for up to 40 years. Trees of all sizes above 10 cm DBH die, are replaced and grow at about 1-2 % annually, but slow-growing trees are more likely to die than fast-growing trees. Growth of an individual tree is much more predictable than the population mean because it is strongly correlated with past growth rate; tree size tends to increase linearly over long periods. These effects mean that fast-growing trees, are more likely to reach maturity than slow-growing trees, and that they can do so in a relatively short time. Increased crown illumination, after logging or natural events leading to canopy openings, causes increased growth but may also lead to higher mortality rates. These effects are detectable for all guilds of tree species in tropical forest, even in understory trees which may spend their whole life in shade.
Significant changes in species population size have been found, but even 40 years is only a small part of the life of most forest trees. Conclusions about compositional equilibrium (or lack of it) over such short periods can only be speculative. Of more interest are the differences we can detect between species in the parameters of population dynamics (fecundity, recruitment, growth, and mortality) and in their responses to environmental influences (especially light, moisture, and nutrients). At present, our knowledge of individual species ecology is uneven and very incomplete, but is much needed if we are to develop management methods for natural tropical forest. Without such knowledge we will be unable to deliver a sustainable and predictable yield from forests, and ecologists and silviculturalists will be regarded as self-indulgent and irrelevant. Tropical forests will then be converted to low-diversity systems for management by economists and engineers (*M. D. Swaine, University of Aberdeen, St. Machar Drive, Aberdeen AB9 2UD, Scotland, United Kingdom*)

POSTER

Distribution of tree species of an Amazonian rain forest

K. Thomsen

Species composition was recorded for three samples of primary rain forest in Añangu, Amazonian Ecuador:
1) A 2 km long transect of floodplain forest 4 to 5 m above river level; 2) a 4 km long transect (1.1 ha) of terra firme forest running between 15 and 120 m above river level; 3) 1.0 ha of terra firme forest 115 to 125 m above river level.
The samples comprised 1954 trees of 10 cm DBH or more, and 396 species. The distributional ranges were analyzed for 305 of the species.
The floodplain and the high terra firme samples had only few species in common, but both had a number of species in common with the line transect. Species confined to terra firme which were more common in the high terra firme sample than in the transect, occurred 4.8 times more often in the upper half of the transect than in the lower half. Of these species, 82.4 % were widely distributed in Amazonia. These species included *Quararibea ochrocalyx*, *Aparisthmium cordatum*, and *Virola elongata*. The species confined to the floodplain and the transect (at least two individuals) occurred 3.3 times more often in the lower half of the transect than in the upper half. Only 50.0 % of these species were widely distributed in Amazonia. The species included *Bauhinia arborea*, *Pentagonia macrophylla*, and *Rinorea apiculata*. Among the species that occurred in all three samples, 66.3 % were widely distributed in Amazonia.
Of the species restricted to the high terra firme sample, 27.0 % were also found in southeastern Brazil. For the species restricted to the floodplain sample, this value was only 6.6 % and for all the species it was 19.7 %. *(K. Thomsen, Botanical Institute, University of Aarhus, Nordlandsvej 68, DK-8240, Risskov, Denmark)*

POSTER

Ink drawing and botanical illustration

K. Tind

Staff, students and visiting botanists at the Botanical Institute, University of Aarhus prepare illustrations in collaboration with artist Kirsten Tind and technician Anni Sloth. The poster shows how illustrations can give much more information than several written pages. Large plants as well as microscopical characters can be fitted into the size of a journal page. The poster shows an illustration of a palm tree. This ink drawing combines information from herbarium specimens, pickled material, slides, and photographs. *(K. Tind, Botanical Institute, University of Aarhus, Nordlandsvej 68, DK-8240 Risskov, Denmark)*

POSTER

Do tree seedlings do better in gaps?: some observations from a Malaysian coastal hill Dipterocarp rain forest

I. Turner

Observations on the growth and survival of tree seedlings were made in Pantai Aceh Forest Reserve, Penang Peninsular Malaysia. Four natural gaps and two closed-canopy sites were used. In each site twenty 1 m^2 plots were laid out at random and all seedlings in them were tagged, mapped and measured for height, basal stem diameter and number of leaves. After 18 months, seedlings were reenumerated and remeasured. A hemispherical canopy photograph was taken for each 1 m^2 plot. These have been analysed with a computerized system to calculate the diffuse site factor for each plot.
Analyses to date show no significant relationship between the radiation regime of a plot, as measured by the diffuse site factor, and the seedling mortality and growth. Possible reasons for this are discussed. (*I. Turner, Oxford Forestry Institute, South Parks Road, Oxford OX1 3RB, England, United Kingdom*)

POSTER

Pollination, seed production, and regeneration of *Acacia nilotica*

K. Tybirk

In an attempt to relate seed production and regeneration of African *Acacias* to pollination syndromes, field studies were initiated in Kenya in 1987.
Acacia nilotica has no disc and the inflorescence functions as one "pollen flower" for the pollinator. Only a rather small proportion of the flowers are hermaphroditic, the rest produce pollen for the attraction of bees. Pollinators are solitary pollen collecting bees. *Acacia nilotica* sheds pollen in the morning in 16 grain polyads fitting the stigma cup. With 16 ovules per ovary this seems a perfect adaptation to ensure full seed set following a single pollination event. Only a small proportion of the flowers, however, result in pod and seed set. The ratio of staminate to hermaphrodite flowers and the lack of polyad transfer by bees restrict pod set, while viability of pollen grains and self incompatibility reduce the number of seeds per pod.
Indehiscent pods have evolved in some African *Acacias* as an adaptation to dispersal by large mammals. These species have large rounded seeds with a very hard seed coat to withstand chewing and digestion by the animals. Ingestion of seeds may also enhance germination. Fire followed by rains can initiate

germination of parts of the *Acacia* seed pool in savanna soils. The seeds are, however, rather susceptible to attack by seed boring beetles of the family Bruchidae. The bruchids are capable of destroying a large proportion of the seeds if the pods are not ingested and the seeds are spread by game or domesticated animals. (*K. Tybirk, Botanical Institute, University of Aarhus, Nordlandsvej 68, DK-8240 Risskov, Denmark*)

POSTER

Analysis of the Andean forests of Ecuador at the generic level

C. Ulloa

This work presents information on trees and shrubs native to the Andean forests above 2,400 m a.s.l. To achieve this goal, a taxonomic study with notes on the distribution and a key for generic identification is being completed. The study is based on existing herbarium specimens from the QCA herbarium and new collections made since May 1986 throughout the Andean zone. The systematic treatment includes a description of each family and each genus represented in the study zone.

Preliminary results indicate the presence of 66 families and 151 genera of woody plants. Tree families are characteristically represented by only a few genera. Dominating genera are *Polylepis*, *Miconia* and *Cedrela*. A few families of shrubs are represented by many genera. These include Asteraceae, Ericaceae, and Melastomataceae. (*C. Ulloa, Herbario QCA, Departamento de Ciencias Biológicas, Pontificia Universidad Católica del Ecuador, Apartado 2184, Quito , Ecuador*)

POSTER and DEMONSTRATION

Computer and Rijksherbarium

P. C. van Welzen

Nowadays, a significant part of systematic work can efficiently be automatized. Commercially available programs can be used to perform most of the work, e.g., wordprocessing (Wordperfect, Wordstar 2000), database handling (dBase III+, Delta), key/description construction (Delta), phylogeny reconstruction (Paup, Cafca) and phenetics (Ntsys). However, it pays off to develop special programs for several more specific, often laborious routine tasks. At the Rijksherbarium three of these programs are now in progress:

Herbut (Herbarium utilities), which combines three tasks:
1. It prepares identification labels, with the name of the family, species, date, institute, and identifier.
2. It produces an identification list from a database (Blumea format).
3. A literature database system will be created which can automatically generate the nomenclature headings above the species descriptions (Blumea format).
Coor. This program contains several databases with the coordinates of the collecting localities within the Malesian region. Records can be added, viewed, changed, deleted, sorted, selected, and printed. In the future it will be possible to mark the records per taxon, whereby the tagged records can be used to prepare distribution maps with the aid of the:
Map Drawing Program. One chooses taxa, region and means of output. The collection localities will be displayed as dots on the chosen map.
All three programs are menu-driven. Herbut and Coor are programmed with dBase III+ and the Map drawing program in APL. Examples are taken from the genus *Guioa* (Sapindaceae), presently under revision by the author. (*P. C. van Welzen, Rijksherbarium, Rapenburg 70-74, 2300 RA Leiden The Netherlands*)

POSTER

Flora of upper montane rain forests in Sri Lanka and south India

W. L. Werner

The floristic composition of the upper montane rain forest of Sri Lanka is very peculiar, as many characteristic plants of montane forests in tropical Asia do not occur. All conifers and Fagaceae (*Lithocarpus*, *Castanopsis*) are lacking. The most prominent genus in these forests regarding frequency and tree size is *Calophyllum* (Clusiaceae). Next to *Calophyllum* come Myrtaceae, Lauraceae, Theaceae and various species of *Symplocos* or *Elaeocarpus*. Many of these plants are common in south Indian mountains or even in the Himalayas (e.g., *Michelia nilagirica*, *Ilex wightiana*, *Berberis aristata* and *Rhododendron arboreum*).
In the mountains of south India some northern elements exist, e.g., *Mahonia*, but tropical genera like *Calophyllum* are not represented in the upper montane rain forests. The ecological niche of *Calophyllum* is occupied by *Syzygium calophyllifolium*, which resembles the *Calophyllum* from Ceylon not only in stature, but also in leaf shape.
The mountain floras of Sri Lanka and south India must have evolved together for some time, being isolated from those of SE-Asia, and must be regarded as related plant communities. (*W. L. Werner, South-Asia Institute, Dep. of Geography, Univ. Heidelberg, Im Neuenheimer Feld 330, D-6900 Heidelberg, FRG*)

POSTER

Inventory of the flora and vegetation of the Parque Nacional Podocarpus in Ecuador

B. Øllgaard and J. E. Madsen

Parque Nacional Podocarpus is located in the provinces of Loja and Zamora-Chinchipe, southern Ecuador, across the eastern Cordillera, at an altitude of 1,500-3,500 m. Most of the park area consists of wet montane forest. Smaller areas above the forest limit, alt. ca. 3,000 m, are covered by herbaceous vegetation with few grasses, and were until recently very rarely or not at all influenced by fire. Phytogeographically the area is closer to northern Peru than to northern Ecuador.

A general floristic survey of the park will be carried out during 1988 and 1989.

A sample plot of one hectare of forest at alt. 2,900 m is being established. Two additional plots at alt. ca 2,500 m and ca. 2,200 m will be established during 1988. All trees of 5 cm DBH or more are permanently tagged, mapped, measured, and vouchered, and their phenology will be studied for two years.

In collaboration with Universidad Nacional de Loja, Ministerio de Agricultura y Ganaderia, and Subcomisión Ecuatoriana PREDESUR, a selection of native tree species from the provinces of Loja and Zamora-Chinchipe will be taken into experimental cultivation in existing plant nurseries. Their qualities and requirements for artificial propagation will be studied and the results will be used to increase the number of native trees which can be used in local reforestation programs. Emphasis is placed on the selection of valuable timber trees, sources of which are being rapidly depleted throughout the region. In addition, tree species are selected for reforestation of degraded grassland in the drought-plagued valleys south of Loja. (*B. Øllgaard and J. E. Madsen, Botanical Institute, University of Aarhus, Nordlandsvej 68, DK-8240 Risskov, Denmark*)

PARTICIPANTS

Alvarez, Ricardo G.
Real Jardin Botánico
Plaza de Murillo 2
28014 Madrid, *Spain*

Andersson, Lennart
University of Göteborg
Carl Skottsbergs Gata 22
S-413 19, Göteborg, *Sweden*

Arcos-Terán, Laura
Dept. de Biología, P.U.C.E.
Apart. 2184, Quito, *Ecuador*

Ashton, Peter S.
The Arnold Arboretum of Harvard University,
22 Divinity Avenue, Cambridge,
Massachusetts 02138, *USA*

Astholm, Fanny
University of Göteborg
Carl Skottsbergs Gata 22
S-413 19, Göteborg, *Sweden*

Balslev, Henrik
Bot. Inst., University of Aarhus,
Nordlandsvej 68,
DK-8240 Risskov, *Denmark*

Barfod, Anders
Bot. Inst., University of Aarhus,
Nordlandsvej 68,
DK-8240 Risskov, *Denmark*

Barthelemy, Daniel
Laboratoire de Botanique
163, rue Auguste Broussonet
34000 Montpellier, *France*

Bergman, Birgitte
Bot. Inst., University of Aarhus,
Nordlandsvej 68,
DK-8240 Risskov, *Denmark*

Bernal, Rodrigo G.
Universidad Nacional
Inst. de Ciencias Naturales
A.A. 7495, Bogotá, *Colombia*

Bloch, Klaus
Bot. Inst., University of Aarhus,
Nordlandsvej 68,
DK-8240 Risskov, *Denmark*

Blicher, Ulla
Bot. Inst., University of Aarhus,
Nordlandsvej 68,
DK-8240 Risskov, *Denmark*

Borchsenius, Finn
Bot. Inst., University of Aarhus,
Nordlandsvej 68,
DK-8240 Risskov, *Denmark*

Brandbyge, John
Bot. Inst., University of Aarhus,
Nordlandsvej 68,
DK-8240 Risskov, *Denmark*

Bruenig, E. F.
Ordinat für Weltforstwirtschaft
Leuschnerstrasse 91
D-2050 Hamburg 80, *FRG*

Carlsson, Mats
University of Göteborg
Carl Skottsbergs Gata 22
S-413 19, Göteborg, *Sweden*

Casas, J. Fernández
Real Jardin Botánico
Plaza de Murillo 2
28014 Madrid, *Spain*

Castroviejo, Santiago
Real Jardin Botánico
Plaza de Murillo 2
28014 Madrid, *Spain*

Chantaranothai, P.
School of Botany
Trinity College
Dublin 2, *Ireland*

Chen Zhong-yi
South China Institute of Botany
Guangzhou, Wushan
Peoples Republic of China

Christensen, Henning
Bot. Inst., University of Aarhus,
Nordlandsvej 68,
DK-8240 Risskov, *Denmark*

Coello, Flavio
Min. de Agric. y Ganaderia
Dept. de P. Nac. y Vida Silvestre,
Quito, *Ecuador*

Cornelissen, Hans
University of Utrecht, Heidelberg-
laan 2, P.O.Box 80.102, 3508 TC
Utrecht, *The Netherlands*

Dantas, Mario
Dept. of Zoology, South Parks
Road, Oxford OX1 3PS,
England, UK

Dransfield, John
Royal Botanic Gardens
Kew, Richmond, Surrey
TW9 3AB, *England, UK*

Ek, Renske C.
University of Utrecht, Heidelberg-
laan 2, P.O.Box 80.102, 3508 TC
Utrecht, *The Netherlands*

Eliasson, Uno
University of Göteborg
Carl Skottsbergs Gata 22
S-413 19, Göteborg, *Sweden*

Elleman, Lis
Bot. Inst., University of Aarhus,
Nordlandsvej 68,
DK-8240 Risskov, *Denmark*

Eriksen, Bente
University of Göteborg
Carl Skottbergs Gata 22
S-413 19, Göteborg, *Sweden*

Eriksson, Roger
University of Göteborg
Carl Skottsbergs Gata 22
S-413 19, Göteborg, *Sweden*

Feuillet, Christian
Centre ORSTOM de Cayenne
B.P. 165, 97323 Cayenne Cedex,
French Guiana

Foster, Robin B.
Field Mus. of Nat. History
Roosevel Road at Lake Shore
Drive, Chicago, Ill. 60605, *USA*

Fredrikson, Margit
University of Göteborg
Carl Skottsbergs Gata 22
S-413 19, Göteborg, *Sweden*

Gamarra, Roberto
Real Jardin Botánico
Plaza de Murillo 2
28014 Madrid, *Spain*

Garilleti, R.
Real Jardin Botánico
Plaza de Murillo 2
28014 Madrid, *Spain*

Geesink, R.
Rijksherbarium, Rapenburg 70-74,
2311 EZ Leiden, *The Netherlands*

Gentry , Alwin H.
Missouri Botanical Garden
P.O. Box 299, St. Louis,
Missouri 63166-0299, *USA*

Grignon, Isabelle
Bot. Inst., University of Aarhus,
Nordlandsvej 68,
DK-8240 Risskov, *Denmark*

Griffo, F.
Liberty Hyde Bailey Hortorium
Cornell University
467 Mann Library Building
Ithaca, New York 14853, *USA*

Görts-van Rijn, A. R. A.
University of Utrecht, Heidelberglaan 2, P.O.Box 80.102, 3508 TC
Utrecht, *The Netherlands*

Hagberg, Mats
University of Göteborg
Carl Skottsbergs Gata 22
S-413 19, Göteborg, *Sweden*

Hamann, Ole
University of København
Gothersgade 140,
DK-1123 København K.,
Denmark

Hansen, Bertel & Mrs. Hansen
Botanical Museum
Gothersgade 130,
DK-1123 København K.,
Denmark

Harling, Gunnar & Mrs. Harling
University of Göteborg
Carl Skottsbergs Gata 22
S-413 19, Göteborg, *Sweden*

Hartshorn, Gary S.
Tropical Science Center
Apartado 8-3870, San José
Costa Rica

Haynes, Robert R.
Univ. of Alabama, P.O. Box
1927, 35486 Alabama, *USA*

Hedberg, Olov
University of Upsala
P.O. Box 541,
S-75 121 Upsala, *Sweden*

Holm-Nielsen, Lauritz B.
Danish Research Academy
Palludan Müllers Vej 17
DK-8000 Aarhus C., *Denmark*

Hu Chi-ming
South China Institute of Botany
Guangzhou, Wushan
Peoples Republic of China

Huber, Otto & Mrs. Huber
C.V.G. and Instituto Venezolano
de Investigaciones Cientificas
Apartado 80405,
Caracas 1080-A, *Venezuela*

Irion, Georg
Senckenberg Institut
Schleussenstrasse 39A,
D-2940 Wilhelmshaven, *FRG*

Iwatsuki, Kunio
University of Tokyo Botanic
Gardens, 3-7-1 Hakusan
Tokyo 112, *Japan*

Jansen-Jacobs, M. J.
University of Utrecht, Heidelberg-
laan 2, P.O.Box 80.102, 3508 TC
Utrecht, *The Netherlands*

Jormalainen, Veijo
University of Turku
SF-20500 Turku, *Finland*

Junk, Wolfgang J.
Max-Planck Institut für Limnologie
Postfach 165 D-2320 Plön, *FRG*

Jørgensen, Peter Møller
Dept. de Biología, P.U.C.E.
Apart. 2184, Quito, *Ecuador*

Kalkman, C.
Rijksherbarium, Rapenburg 70-74,
2300 RA Leiden, *The Netherlands*

Kalliola, Risto
University of Turku
SF-20500 Turku, *Finland*

Klitgaard, Bente Bang
Bot. Inst., University of Aarhus,
Nordlandsvej 68,
DK-8240 Risskov, *Denmark*

Knudsen, Jette Teilman
University of Göteborg
Carl Skottsbergs Gata 22
S-413 19, Göteborg, *Sweden*

Korning, Jørgen
Bot. Inst., University of Aarhus,
Nordlandsvej 68,
DK-8240 Risskov, *Denmark*

Köhler, Egon
Humboldt-Universität zu Berlin
Späthstr. 80/81
DDR-1195 Berlin, *GDR*

Kullberg, Eva
Marselisborg Gymnasium
DK-8000 Århus C., *Denmark*

Kvist, Lars Peter
Bot. Inst., University of Aarhus,
Nordlandsvej 68,
DK-8240 Risskov, *Denmark*

Larsen, Kai
Bot. Inst., University of Aarhus,
Nordlandsvej 68,
DK-8240 Risskov, *Denmark*

Larsen, Supee Saksuwan
Bot. Inst., University of Aarhus,
Nordlandsvej 68,
DK-8240 Risskov, *Denmark*

Lawesson, Jonas E.
Bot. Inst., University of Aarhus,
Nordlandsvej 68,
DK-8240 Risskov, *Denmark*

Lebrón-Luteyn, María
New York Botanical Garden
Bronx, NY 10458, *USA*

Lindström, Marie
University of Göteborg
Carl Skottsbergs Gata 22
S-413 19, Göteborg, *Sweden*

Lundin, Roger
Naturhistoriska Riksmuseet
Box 50007, S-104 05 Stockholm
Sweden

Luteyn, James L.
New York Botanical Garden
Bronx, NY 10458, *USA*

Lægaard, Simon
Bot. Inst., University of Aarhus,
Nordlandsvej 68,
DK-8240 Risskov, *Denmark*

Maas, Paul J. M. & Mrs. Maas
University of Utrecht, Heidelberglaan 2, P.O.Box 80.102, 3508 TC
Utrecht, *The Netherlands*

Madsen, Jens
Bot. Inst., University of Aarhus,
Nordlandsvej 68,
DK-8240 Risskov, *Denmark*

Madsen, Eva B.
Bot. Inst., University of Aarhus,
Nordlandsvej 68,
DK-8240 Risskov, *Denmark*

McMeekin, Douglas
c/o Alex Chapman, CONOCO DU
3104, P.O.Box 2197, Houston,
Texas 77252, *USA*

Molau, Ulf
University of Göteborg
Carl Skottsbergs Gata 22
S-413 19, Göteborg, *Sweden*

Moraes Ramirez, Mónica
Herbario Nacional de Bolivia
Cajón Postal 20127
La Paz, *Bolivia*

Mori, Scott Allan
New York Botanical Garden
Bronx, NY 10458, *USA*

Neill, David
USAID / Quito, AID, Washington
DC 20523, *USA*

Neuendorf, Magnus
University of Göteborg
Carl Skottsbergs Gata 22
S-413 19, Göteborg, *Sweden*

Newmann, Mark
University of Aberdeen
St. Machar Drive, Aberdeen
AB9 2UD, *Scotland, UK*

Nielsen, Ingvar
Bot. Inst., University of Aarhus,
Nordlandsvej 68,
DK-8240 Risskov, *Denmark*

Nielsen, Ivan
Bot. Inst., University of Aarhus,
Nordlandsvej 68,
DK-8240 Risskov, *Denmark*

Nordenstam, Bertil
Naturhistoriska Riksmuseet
Box 50007, S-104 05 Stockholm
Sweden

Oldeman, Roelof A. A.
AUW-B&B, P.O. Box 342,
6700 AN, Wageningen,
The Netherlands

Parnell, J.
School of Botany,
Trinity College
Dublin 2, *Ireland*

Pedersen, Henrik Borgtoft
Bot. Inst., University of Aarhus,
Nordlandsvej 68,
DK-8240 Risskov, *Denmark*

Polhill, R. M.
Royal Botanic Gardens
Kew, Richmond, Surrey
TW9 3AB, *England, UK*

Ponce, Arturo
Min. de Agric. y Ganaderia
Dept. de P. Nac. y Vida Silvestre,
Quito, *Ecuador*

Poulsen, Axel D.
Bot. Inst., University of Aarhus,
Nordlandsvej 68,
DK-8240 Risskov, *Denmark*

Puhakka, Marit
University of Turku
SF-20500 Turku, *Finland*

Rajasilta, M
University of Turko
SF-20500 Turko, *Finland*

Raven, Peter H.
Missouri Botanical Garden
P.O. Box 299, St. Louis,
Missouri 63166-0299, *USA*

Renner, Susanne Sabine
Bot. Inst., University of Aarhus,
Nordlandsvej 68,
DK-8240 Risskov, *Denmark*

Richford, Andrew
Academic Press Limited
24-28 Oval Road, London NW1
7DX, *England, UK*

Riedl, Harald
Vienna Natural History Museum
Dept. of Botany, Burgring 7
A-1014 Vienna, *Austria*

Riedl-Dorn, Christa
Vienna Natural History Museum
Dept. of Botany, Burgring 7
A-1014 Vienna, *Austria*

Ruokalainen, Kalle
University of Helsinki
SF-00100, Helsinki, *Finland*

Saldarriaga, Juan G.
Convenio Colombo - Holandés
Calle 20 No. 5-44,
A.A. 034174, Bogotá , *Colombia*

Salo, Jukka
University of Turku
SF-20500 Turku, *Finland*

Santisuk, T.
The Forest Herbarium
Royal Forest Department
Bangkok 10900, *Thailand*

Schoser, Gustav & Mrs. Schoser
Palmengarten, Siesmayerstrasse 61
D-6000 Frankfurt am Main 1,
FRG

Schmidtbauer, Bernt
Der Deutsche Bundestag
Bundeshaus A 28
D-5300 Bonn 1, *FRG*

Shukla, Vidjai K. S.
Aarhus Olifabrik A/S
M.P. Brunsgade 27
DK-8100 Aarhus C., *Denmark*

Silva, J. N. M.
Oxford Forestry Institute
South Parks Road
Oxford OX1 3RB, *England, UK*

Simonis, J. E.
University of Utrecht, Heidelberg-
laan 2, P.O.Box 80.102, 3508 TC
Utrecht, *The Netherlands*

Sirirugsa, Puangpen
Prince of Songkla University,
Hat Yai 90110, *Thailand*

Skov, Flemming
Bot. Inst., University of Aarhus,
Nordlandsvej 68,
DK-8240 Risskov, *Denmark*

Smitinand, Tem
The Forest Herbarium
Royal Forest Department
Bangkok 10900, *Thailand*

Smith, Lars
Bot. Inst., University of Aarhus,
Nordlandsvej 68,
DK-8240 Risskov, *Denmark*

Smits, Willie & Mrs. Smits
P.O. Box 220
Balikpapan, *Indonesia*

Spichiger, Rodolphe & Mrs. Spichiger
Conservatoire et Jardin Botaniques
Case postale 60, CH-1292
Chambésy/GE, *Switzerland*

Stutz de Ortega, Liliane
Conservatoire et Jardin Botaniques
Case postale 60, CH-1292
Chambésy/GE, *Switzerland*

Ståhl, Bertil
University of Göteborg
Carl Skottsbergs Gata 22
S-413 19, Göteborg, *Sweden*

Sumithraarachchi, D. B.
Royal Botanic Garden
Peradeniya, *Sri Lanka*

Swaine, M. D.
University of Aberdeen
St. Machar Drive, Aberdeen
AB9 2UD, *Scotland, UK*

Thomsen, Karsten
Bot. Inst., University of Aarhus,
Nordlandsvej 68, DK-8240
Risskov, *Denmark*

Thygesen, Anders
Bot. Inst., University of Aarhus,
Nordlandsvej 68,
DK-8240 Risskov, *Denmark*

Tind, Kirsten
Bot. Inst., University of Aarhus,
Nordlandsvej 68,
DK-8240 Risskov, *Denmark*

Tuomista, Hanne
University of Helsinki,
SF-00100, Helsinki, *Finland*

Turner, Ian
Oxford Forestry Institute
South Parks Road
Oxford OX1 3RB, *England, UK*

Tybirk, Knud
Bot. Inst., University of Aarhus,
Nordlandsvej 68,
DK-8240 Risskov, *Denmark*

Ulloa, Maria del Carmen
Dept. de Biología, P.U.C.E.
Apart. 2184, Quito, *Ecuador*

Vidal, J. E.
Laboratoire de Phanérogamie
16 rue Buffon
75005 Paris, *France*

Welle, Ben J. H. ter
University of Utrecht, Heidelberg-
laan 2, P.O.Box 80.102, 3508 TC
Utrecht, *The Netherlands*

Welzen, P. C. van
Rijksherbarium, Rapenburg 70-74,
2300 RA Leiden, *The Netherlands*

Werner, Wolfgang L.
South-Asia Institute, Univ.
Heidelberg
In Neuenheimer Feld 330
D-6900 Heidelberg, *FRG*

Westra, L. Y.
University of Utrecht, Heidelberg-
laan 2, P.O.Box 80.102, 3508 TC
Utrecht, *The Netherlands*

Zinck, Alfred
ITC , P.O Box 6
7500 AA Entschede, *The Netherlands*

Øllgaard, Benjamin
Bot. Inst., University of Aarhus,
Nordlandsvej 68,
DK-8240 Risskov, *Denmark*

SYMPOSIUM STAFF

Brandt, Birthe
Deleuran, Holger
Højstrøm, Britta
Nørgaard, Flemming
Sloth, Annie

Bot. Inst., University of Aarhus,
Nordlandsvej 68,
DK-8240 Risskov, *Denmark*

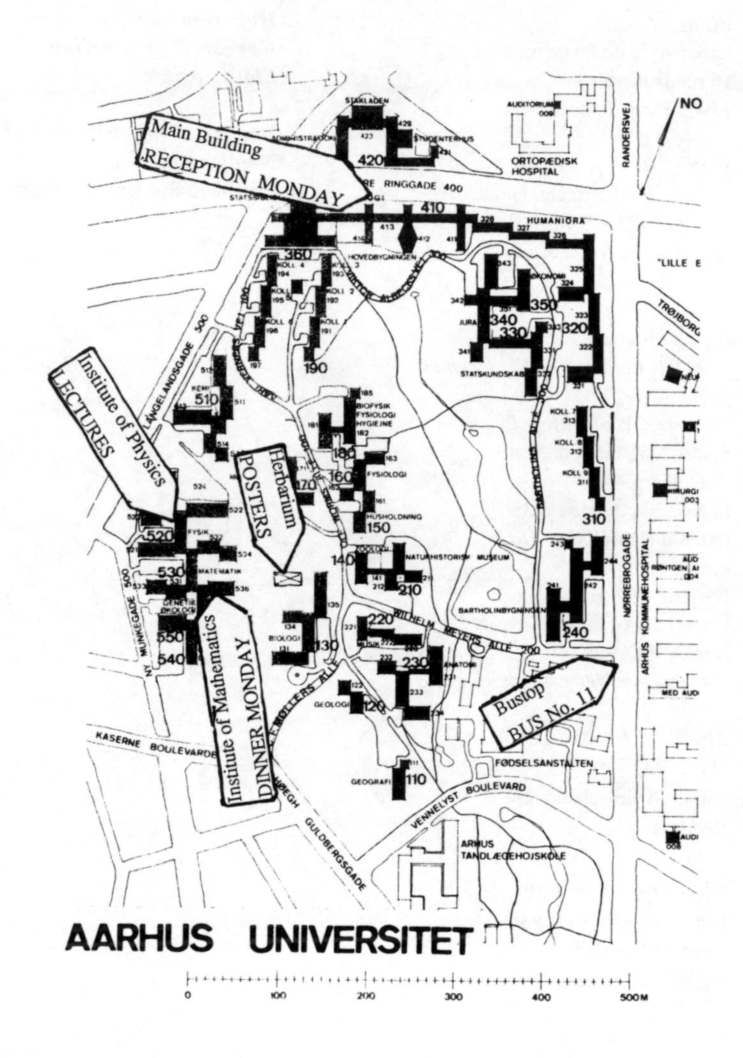
Main Building
RECEPTION MONDAY
Institute of Physics
LECTURES
Herbarium
POSTERS
Institute of Mathematics
DINNER MONDAY
Bustop
BUS No. 11
NO
RANDERSVEJ
ORTOPÆDISK HOSPITAL
HUMANIORA
HOVEDBYGNINGEN
JURA
STATSKUNDSKAB
ØKONOMI
KEMI
FYSIK
MATEMATIK
BIOFYSIK
FYSIOLOGI
HYGIEJNE
HUSHOLDNING
ZOOLOGI
NATURHISTORISK MUSEUM
BARTHOLINBYGNINGEN
WILHELM MEYERS ALLÉ
BIOLOGI
MUSIK
ANATOMI
GEOLOGI
GEOGRAFI
FØDSELSANSTALTEN
VENNELYST BOULEVARD
ÅRHUS TANDLÆGEHØJSKOLE
GULDBERGSGADE
KASERNE BOULEVARDEN
NY MUNKEGADE
LANGELANDSGADE
NØRREBROGADE
ÅRHUS KOMMUNEHOSPITAL
AARHUS UNIVERSITET
0
100
200
300
400
500M